Frogs are amazing animals. They have been around since the Age of the Dinosaurs. They have no fur, no feathers, and no scales. They have lungs—but can also breathe through their skin. Frogs have long, powerful back legs to jump through the air. A frog can jump more than 20 times the length of its body!

A frog has no ribs. This skeleton shows the length of its hind legs and its small upper body.

Many frogs go through metamorphosis (meta-MOR-pho-sis)—the change from tadpoles that live in water into adult frogs that live on land. Animals that can live in water and on land are called amphibians. The word comes from the Greek word *amphibios,* which means "double life."

Many frogs, like the wood frog, go through three life stages. They start as clumps of eggs that female frogs lay in the water. The eggs sink to the bottom of the water. Soon, the eggs drift to the surface of the water.

The eggs hatch into little fishlike creatures with long tails. They are called tadpoles or polliwogs. At first, the tadpole breathes through gills, like a fish.

Soon the tadpole begins to grow legs. Its hind legs appear first.

Then its lungs begin to develop and its front legs emerge.

The tadpole loses its gills. Finally, a tiny froglet, with just the stump of a tail, comes out of the water.

The froglet soon loses its tail. Now it looks like an adult frog.

Although metamorphosis enables the adult frog to breathe air and live on land, most frogs still need water for the moisture it provides their skin. These frogs live around bodies or pools of water where they will mate, deposit their eggs, and continue the life cycle.

Bullfrog: The largest frog in the United States, the bullfrog spends almost all of its time in or around water.

Horned Frog: Fierce and fanged, this South American frog (also called Bell's frog) hops around the jungle floor and hides in dirt or leaves to catch its next meal. The horned frog travels to water only to mate and lay eggs.

Red-Eyed Tree Frog: The littl red-eyed tree frog lives mostly i the pool of water that forms in the center of bromeliad plants i tropical rain forests.

Frog or toad? People often confuse these two amphibians. Although frogs and toads seem alike, they differ in a few ways. Most herpetologists (HER-pe-TOL-o-gists), the scientists who study reptiles and amphibians, use the name "frogs" to mean both frogs and toads.

Frogs spend more time in the water. A frog's skin is moist and smooth. It has a slim body and long legs.

Toads spend more time on land and less time in the water. A toad's skin is thicker and bumpier. It has a plump body and shorter legs.

Frogs are found on every continent, except Antarctica. There are over 3,500 frog species. They vary greatly in color, size, and even foot shape. In fact, the shape of a frog's hand or foot is a good way to tell how a frog moves and where it lives.

Tree frogs are excellent climbers. They use the round sticky pads at the ends of their fingers and toes to grab the branches. This is the barred leaf frog, from Australia.

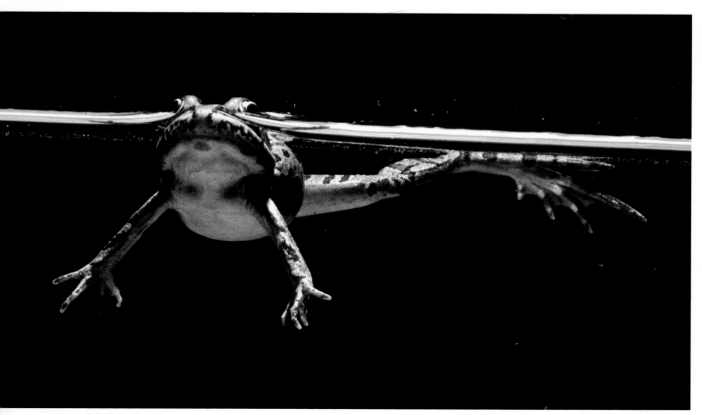

Swimming frogs use their webbed hind feet to speed through the water. Their toes grab on to rocks and weeds, protecting the frogs from powerful currents. This spotted North American swimmer is the leopard frog.

Digging or burrowing frogs use the sharp growths on their hind feet. This Australian banjo frog digs a hiding place in the soil as soon as it senses danger.

But whatever a frog's color, size, or shape, most frogs have some characteristics in common.

A frog's skin is sensitive enough to absorb moisture, heat from the sun, and cold. A frog feels a sudden change in temperature and will move quickly to a sunny spot to get warmer or to a shady spot under the leaves of a tree to get cooler.

Even though a frog has excellent hearing, it doesn't have outer ears. It hears through the round patch behind each eye called a tympanum (TIM-pa-num).

A frog's bulging eyes let it see in almost all directions at once. This helps the frog see its enemies as well as catch its prey. When a frog sees an insect walking or flying by, it quickly flicks out its sticky tongue and grabs it.

Another similarity is that it is usually the male frog that "advertises" its presence with sound. The male frog is able to make this sound without opening its mouth. The female frog hears the calls, but most female frogs do not make these sounds at all.

Can you tell which of these frogs is making sounds?

The bullfrog repeats a low-pitched song that sounds like "jug-o-rum, jug-o-rum."

As its name suggests, the barking tree frog sounds like a dog.

In early spring hundreds of spring peepers sing a high-pitched "peeeeep, peeeeep."

Most frogs have clever ways of protecting themselves from danger.

Believe it or not, the frogs above and below are both gray tree frogs. The gray tree frog can change its color from green to gray, which helps it to blend into its surroundings. This defense is known as *camouflage*.

When confronted with danger, some frogs, like the Asian horned frog, will try to look fierce and make a threatening posture. Other frogs might even puff themselves up to appear bigger than they actually are.

Is this frog coming or going? It's hard to tell, but the four-eyed frog has turned its back and exposed its eyespots. This is meant to startle and confuse the enemy, which may think it is under attack.

Although these frogs look almost identical, the frog on the left is a poison-dart frog and the frog on the right is a harmless leptodactylid (LEP-toe-DAK-til-id). Looking like a poisonous frog has its advantages. Predators mistake the leptodactylid for a poisonous frog and stay away. This defense is called *mimicry*.

Frogs play an important role in the web of life. They snap up huge numbers of insects and worms as well as small fish and other small animals.

At the same time, snakes, turtles, skunks, raccoons, wading birds, and large fish snap up frogs.

By eating large numbers of insects that might otherwise become serious pests, frogs help humans. But frogs also help humans in other ways.

Not only do mosquito bites hurt but some tropical mosquitoes spread disease. Frogs control the mosquito population by eating great amounts of these pests.

Australian tree frogs give off a chemical that helps to heal sores on human skin. Other frogs produce chemicals that can be used as painkillers or may one day even cure diseases.

By studying frogs in their habitats, some scientists believe they can learn how radiation from the sun affects wildlife. The scientists believe that too much harmful radiation may be responsible for the destruction of frog eggs in wetlands.

But today, some frog species are disappearing. Some may even be extinct. One reason is that when cities and farmlands expand, people destroy frog habitats by cutting down trees and draining wetlands. Also, chemicals used to control insects and weeds pollute the waters in which frogs live. Whatever the reasons, the fact remains. Some species are disappearing.

Red-legged frogs, once a common species of the United States, can't be found in their usual habitats.

Golden mantellas live only in Madagascar. But their habitats are being destroyed, and today very few remain.

Harlequin frogs used to be found in the mountains of South America, but they have not been seen since the late 1980s.

The disappearance of the California Cascades frogs is still a mystery. So far there is no evidence of water pollution in the Rocky Mountain regions where the frogs were once found.

Northern leopard frogs were once the most common in North America. But in Canada they are dying out and have almost disappeared.

There used to be many tiny, beautiful glass frogs in Costa Rica. Today, few glass frogs can be found.

Frogs have been around since the dinosaurs. But now frogs need our help. What can we do to make sure they continue to survive?

Adopting a frog pond is one way to help. By keeping a close watch on frogs and other pond wildlife, we can take action to preserve the natural habitat.

Index

03
ISBN: 1-56784-23